Llyfrgell Dyfed Library

Dosb./Class	Rhif/Acc. No
J574.543 DAV	
Llyfrwerthwr/Supplier PETERS	Rhif Ant./Inv. No. 087936

Time and Change

Kay Davies
and
Wendy Oldfield

Wayland

Starting Science

Books in the series

Animals
Day and Night
Electricity and Magnetism
Floating and Sinking
Food
Hot and Cold
Information Technology
Light
Local Ecology
Materials

Plants
Pushing and Pulling
Rocks and Soil
The Senses
Skeletons and Movement
Sound and Music
Time and Change
Waste
Water
Weather

About this book

Time and Change focuses on topics which closely reflect our changing seasons. They are carefully chosen to be within the child's early experiences of the natural world. Children will learn how the changes in length of daylight and weather that affect our lives, also affect the life cycle of plants and animals.

The activities and investigations in this book are designed to be straightforward but fun, and flexible according to the abilities of the children. With adult guidance they will be introduced to methods of scientific enquiry and recording.

The main picture and its commentary may be taken as a focal point for further discussion or as an introduction to the topic. Each chapter can form a basis for extended topic work.

Teachers will find that in using this book, they are reinforcing the other core subjects of language and mathematics. Through its topic approach ***Time and Change*** covers aspects of the National Science Curriculum for key stage 1 (levels 1 to 3).

First published in 1992 by
Wayland (Publishers) Ltd
61 Western Road, Hove
East Sussex, BN3 1JD, England

© Copyright 1992 Wayland (Publishers) Ltd

Typeset by Kalligraphic Design Ltd, Horley, Surrey
Printed in Italy by
   Rotolito Lombarda S.p.A., Milan
Bound in Belgium by Casterman S.A.

British Library Cataloguing in Publication Data
Davies, Kay
   Time and Change. – (Starting Science)
   I. Title.   II. Oldfield, Wendy   III. Series
   372.3

ISBN 1 85210 997 1

Editor: Mandy Suhr
Series editor: Cally Chambers

CONTENTS

Four seasons	5
Harvest festival	6
Coat of many colours	9
Silken threads	11
Mushroom magic	12
Eating and sleeping	15
Buried treasure	16
Buds and blossoms	18
New life	21
Water ways	22
Eating machine	24
Feeding time	27
Summer goodbye	28
Glossary	30
Finding out more	31
Index	32

All the words that appear in **bold** in the text are explained in the glossary.

These pictures are all of the same place. Can you see the changes throughout the year?

FOUR SEASONS

We can divide our year into spring, summer, autumn and winter. We call these the seasons.

As the seasons change so do the lengths of night and day. Each season has its own weather.

Because of the changing seasons, plants can grow, make fruit and **seeds**, and then rest.
Around the year the lives of creatures change with the seasons.

Decide which of these clothes would be best for spring, summer, autumn or winter.

What do you notice about your clothes for spring and autumn?

HARVEST FESTIVAL

In late summer and early autumn, many fruits and berries, nuts and grains are ready to be harvested.

Fruits like plums and apples can be picked. Hazelnuts and sweet chestnuts are ripe.

Wheat is cut to be ground into flour for our bread and cakes.

Fruits, berries and nuts have seeds inside. The seeds may grow into new plants in spring.

Collect different kinds of seeds and nuts.

You can make pictures and patterns.

Some seeds will need to be dried before you use them.

There are plenty of ripe berries for birds and other creatures to eat in autumn.

In autumn, some leaves begin to change colour and die. They fall to the ground before winter comes.

COAT OF MANY COLOURS

Deciduous trees must lose their leaves before winter. Strong winds blow against the leaves.
Snow settles on them and makes them heavy.
This can break the branches and damage the tree.

Collect fallen leaves. Paint one side of the leaf in an autumn colour. Press it firmly on to paper to make a print.

Fir
Privet
Holly
Laurel

Evergreen trees are special because they keep their leaves all year round.

Their leaves are often thin to let the wind blow between them.

Some leaves are fat and shiny to let the snow slip off.

Drops of autumn **dew** cling to spiders' webs. They sparkle like necklaces in the early morning sunshine.

SILKEN THREADS

The garden spider dies in winter. There are no flies to eat.

She hides her sack of eggs under a shed roof or on a fence.

In the spring, hundreds of spiderlings will hatch. They drift in the air on silken threads.

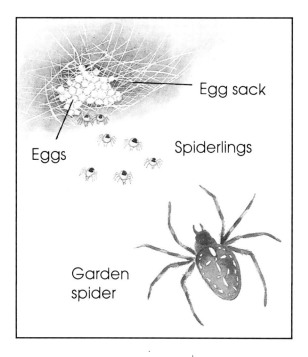

Use wool and glue to make a spider's web on black paper.

Make the spokes first, then join them with circles of wool.

Use four pipe cleaners and crumpled tissue paper to make a spider.

Hang it on your web.

MUSHROOM MAGIC

Many mushrooms and other **fungi** need the wet, warm weather of early autumn to grow.

Their tiny seeds are called **spores**. They are so light that they can be blown by the wind.

A pile of fallen leaves is a good place for the spores to land.
They may grow into new mushrooms.

Leave a mushroom **cap** on some paper overnight.

Lift it up the next day. You will see millions of spores on the paper showing the shape of the cap. Don't speak or they will be carried away by your breath.
Always wash your hands after touching mushrooms.

Autumn is a good time for spotting mushrooms and toadstools. Ask an adult before you touch them as they can be poisonous.

This squirrel is lucky. It has found a nut to eat on the cold, hard, winter ground.

EATING AND SLEEPING

Some creatures must be ready when winter comes. There will be no nuts, berries or insects to eat.

Squirrels keep stores of nuts in cracks in the ground or trees. They hope to have enough to last through the winter. They rest a lot to save energy.

Some creatures don't eat in winter at all.
The dormouse eats lots of nuts in autumn and becomes very fat. It sleeps through the winter.

This sleep is called **hibernation**.

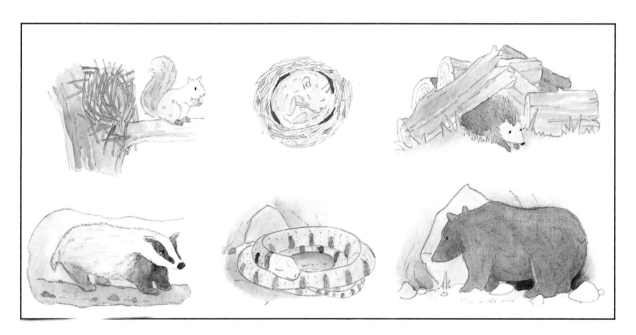

These creatures hibernate.

Can you find out their names?

BURIED TREASURE

Bulbs are food stores.
A new plant will use the food to grow from inside.

In spring, the weather gets warmer. **Roots** grow in the soil from each bulb.

We see **stems**, leaves and flowers growing above the ground.

Rest a hyacinth bulb on top of a jar of water.
Leave it in the light.
Look at it every day.

Where do the roots grow from?
What colour are they?

Watch for the stem beginning to grow.
What colour do you think it will be?

These flowers grew from bulbs buried in the soil. Their bright colours tell us it is springtime.

BUDS AND BLOSSOMS

Deciduous trees are bare in winter.

In spring, look closely at the branches. You will see **buds** wrapped tightly in waterproof cases. Inside are leaves or flowers waiting to grow.

Some trees grow their flowers first. Others grow their leaves first.

Collect twigs from trees in spring. Keep them in water. Watch the buds open and the leaves or flowers uncurl.

The cherry tree has lovely pink **blossom**. It is easy to see.

Look hard at sycamore trees to see their small greenish flowers.

When warmer weather comes in spring, the buds on the trees burst open. New leaves and blossom grow quickly.

These baby birds have hatched from eggs laid in early spring. They are too young to fly or feed themselves.

NEW LIFE

We have more hours of daylight in spring. Birds are busy building nests to lay their eggs.
When the eggs hatch in late spring there will be many insects for the babies to eat.

Other creatures are starting their lives too.
It is a good time for farm animals to be born.
The spring's warmth and rain makes the grass grow.

Some animal mothers must eat plenty of grass. It helps them make milk for their young.

Which of these babies are usually born in spring when there is lots of grass to eat?
Which animals don't live on grass and so can be fed whenever they are born?

WATER WAYS

Fish and water snails live in ponds all year round. In warm weather many creatures use the pond as a nursery for their eggs and babies.

In spring you might see frog spawn or newt spawn. Later, you will find tadpoles and the **larvae** of many sorts of insects.

Plants grow in the water too. They are food for many creatures that live there.

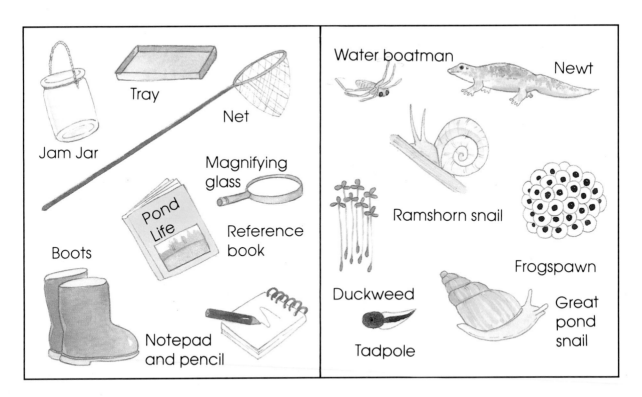

Collect these things and go pond dipping with an adult.
Draw some of the tiny things you find.
Always put everything you catch back in the pond.

The spring sunshine melts the ice and warms the water in the pond. Many insects and other creatures will lay their eggs in the warmer water.

EATING MACHINE

Butterflies and moths lay eggs on plants. Caterpillars hatch from the eggs.

Each caterpillar spends the summer days eating. When it is fully grown it crawls to a safe place. Its skin becomes a hard, crusty shell. It is now a **pupa**. Inside, the caterpillar is changing into a butterfly. These three pictures show a butterfly breaking out of its shell.

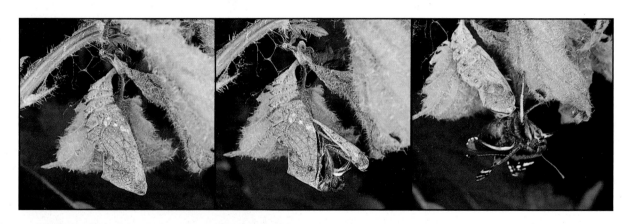

In summer, find a caterpillar to look after. Use a plastic tank with plenty of air holes. Always feed your caterpillar on its **food plant**.

Watch the changes carefully.
Be ready to let your butterfly go at once.

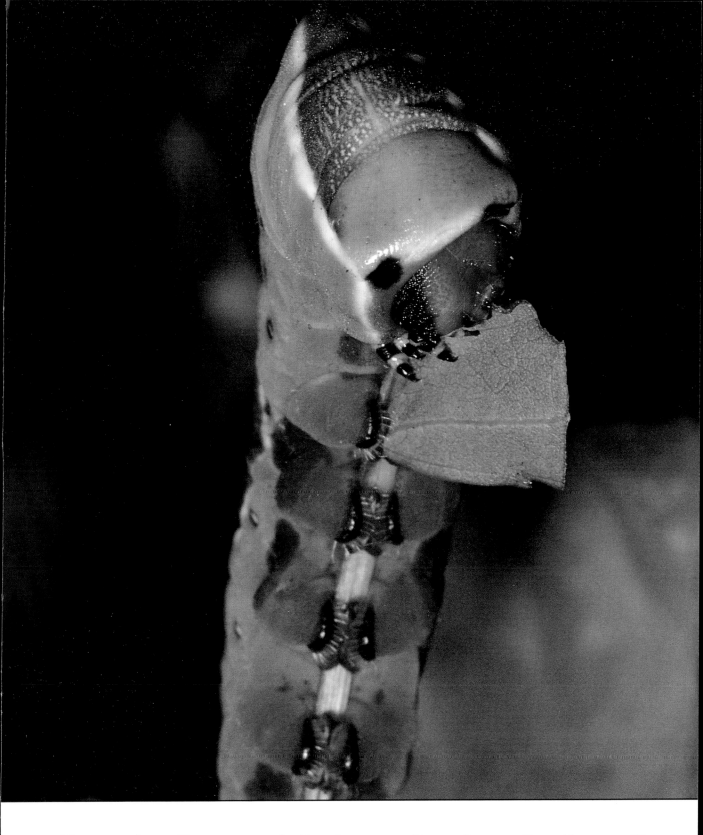

The caterpillar spends its days eating. Soon the time will come for it to change into a butterfly.

The butterfly drinks the sweet nectar from the summer flowers. This is the only food it eats.

FEEDING TIME

In summer, insects feed on **nectar** from flowers. Some insects cannot live in winter when there are no flowers.

Pollen from flowers sticks to the bodies of the insects. Then the pollen is carried to other flowers where it will make fruits and seeds grow.

Make flowers from card and the cups from egg boxes. Paint them bright colours. Stick them on to a board. Dissolve some sugar in water and soak cotton wool in the mixture.
Put the cotton wool into the cups.
Put your flowers outside in the sun.

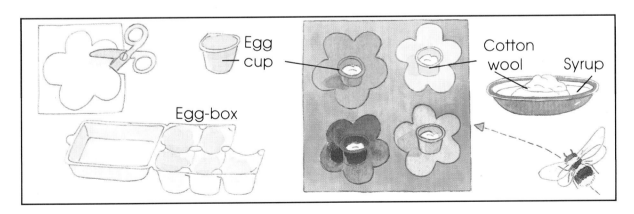

What visits your flowers?
Which colour attracts the most insects?
Why can't you do this in winter?

SUMMER GOODBYE

Many birds live in one country all year round.

Some birds spend only the summer in a country. Before winter comes they must **migrate** because there will be no insects to eat.

They fly thousands of kilometres to find summer weather somewhere else.

Keep a bird diary.

Write down the name of each bird you see and the date.

Which birds do you only see in summer?

Which birds are here just for winter?

At the end of summer the swallows all gather together. Soon they will fly away to warmer lands.

GLOSSARY

Blossom The flowers of a tree or plant.
Buds The growths on stems which contain new leaves or flowers.
Bulbs Swollen underground stems containing food and the shoots of new plants.
Cap The part of a toadstool which contains the spores.
Deciduous trees Trees that lose their leaves in winter.
Dew Drops of water from the air which can be found outside in the early morning.
Evergreen trees Plants which keep their leaves all year.
Food plant A particular plant a creature feeds on.
Fungi Plants that do not have leaves or flowers.
Hibernation Passing the winter in deep sleep.
Larvae Young insects after they have hatched from an egg and before they grow into a pupa and then an adult.
Migrate When birds or animals move to another part of the world for a season.
Nectar The sweet liquid produced by flowers.
Pollen The fine powder produced by plants.
Pupa The stage in an insect's life between larva and adult.
Roots The parts of a plant which take up water and food from the soil.
Seeds The small parts of a plant which can grow if they are planted.
Spores The tiny seeds of a toadstool or mushroom.
Stem The part of a plant from which the leaves, flowers and buds grow.

FINDING OUT MORE

Books to read:

The Garden Year by Christa Spangenberg (A. & C. Black, 1986)
The Orchard by Vanessa Luff (A. & C. Black, 1990)
Snow is Falling by Franklyn M. Branley (A. & C. Black, 1989)

The following series may also be useful:

Starting Points – Autumn/Spring/Summer/Winter by Ruth Thompson (Franklin Watts, 1990)
Through the Seasons – Field and Hedgerow/Garden/Pond/Wood (Wayland Publishers Ltd, 1989)

PICTURE ACKNOWLEDGEMENTS

Bruce Coleman Ltd. 7 (Paton), 12 top (Clement), 13 (Reinhard), 14 (Burton), 17 (Reinhard), 19 (Crichton), 20 (Wilmshurst), 20 (Dore), 24 middle (Burton), 28 top (Wilmshurst), 29 (Meitz); Eye Ubiquitous (Seheult) 8, 18 top and bottom; Frank Lane Picture Agency 29; Oxford Scientific Films 10 (Reinhard), 27 (Bernard), 26 (Maclean); Papilio 25 (Pickett); Wayland Picture Library (Zul Mukhida) cover, 6, 9, 11, 12 bottom, 16, 24 bottom, 28 bottom.

Artwork illustrations by Rebecca Archer.
The publishers would also like to thank Stacey, Terry, Liam, Sade, Chloe and Matthew.

INDEX

Autumn 5, 6, 7, 9, 10, 12, 13, 15

Babies 11, 21, 22
Berries 6, 15
Birds 19, 21
Blossom 18
Branches 9, 18
Buds 18, 19
Bulbs 16, 17
Butterflies 24, 26

Caterpillar 24, 25

Eggs 11, 19, 21, 22, 24

Flowers 16, 17, 18, 19, 26, 27
Food 22, 24, 26, 28
Fruit 5, 6, 7, 27

Hibernation 15
Hyacinth 16

Insects 15, 21, 22, 27, 28

Larvae 22
Leaves 8, 9, 12, 16, 18, 19

Migration 28
Mushroom 12

Nectar 26, 27
Nuts 6, 14, 15

Plants 5, 6, 16, 22, 24
Pollen 27
Pond 22, 23
Pupa 24

Root 16

Seasons 5
Seeds 5, 6, 12, 27
Soil 16, 17
Spiders 10, 11
Spring 5, 11, 16, 17, 18, 19, 21, 22, 23
Stem 16
Summer 5, 6, 24, 27, 28

Tree 19
   cherry 18
   deciduous 9, 18
   evergreen 9
   sycamore 18

Weather 5, 16, 19, 22, 28
Web 10, 11
Winter 5, 8, 9, 11, 14